Chemical Engineering & the Law Forum 2023

Held at the 2023 AIChE Annual Meeting

Orlando, Florida, USA
5-10 November 2023

ISBN: 978-1-7138-9287-8

Printed from e-media with permission by:

Curran Associates, Inc.
57 Morehouse Lane
Red Hook, NY 12571

Some format issues inherent in the e-media version may also appear in this print version.

Printed with permission by Curran Associates, Inc. (2024)

For permission requests, please contact AIChE
at the address below.

AIChE
120 Wall Street, FL 23
New York, NY 10005-4020

Phone: (800) 242-4363
Fax: (203) 775-5177

www.aiche.org

Additional copies of this publication are available from:

Curran Associates, Inc.
57 Morehouse Lane
Red Hook, NY 12571 USA
Phone: 845-758-0400
Fax: 845-758-2633
Email: curran@proceedings.com
Web: www.proceedings.com

TABLE OF CONTENTS

Asset Integrity and Reliability Implementation Challenges and Failure Factors in Developing Countries

Hany Tawfek Sr.

For decades, Asset Integrity Management (AIM) activities have been a part of industry's efforts to prevent incidents and maintain productivity. The study utilized variant references to define the most common challenges and failures of AIM, in advance the study illustrated the activities, workflows, and lifecycle of an AIM, the purpose is to systematically address the correspondent challenges and failures and to avoid overlooking a major challenge or failure factor, then ultimately highlight the possible solutions and recommendations.

The technical data were gathered from the most well-known and globally used guidelines and standards. field data were gathered from interviews with knowledgeable personnel, publishment and research papers, global platforms, and forums and Questionnaire.

The study results and conclusion are concluded mainly from expert reviews, technical papers, technical books, forums, and webinars. The study supported the results with, 1) case study that addressed the AIM implementation failures in 10 production facilities belongs to one of the biggest industrial sectors in Egypt, 2) Questionnaire for an experts and knowledgeable people in the field of Asset integrity in Egypt. Finally, the study addressed a recommendation in correspondent to each failure category.

Failure to apply asset integrity management system practices will definetly increase the probability a major accident to exist, considering the PDCA cycle that represents a typical management system components, failures related to system planning could be avoided by building the AIM needs on a business case, regulatory enforcement that should be supported by spreading the culture of sustainability.

On the other hand, failures related to system implementation could be avoided by defining the Asset integrity program at the top levels in the company, assigning clear responsibilities, and empower implementation.

Failures related to performance assurance could be avoided by Implementation of an encompassing corporate Asset integrityprogram at each local establishment, metrics must be established to evaluate the intended compliance of the corporate MI program.

And lastly for human factors failures, AI responsibilities to be obligated in a dedicated law same as HSE in the Egyptian labor law, availing AI required resources including manning and to be linked to the facilities objectives & obligations (if considered), and train and familiarize the assigned team for AIM implementation with change management function as a key of success.

Patent Litigation in Oil and Gas and Alternative Energy and Its Place in the Broader NPE Litigation Landscape

Kaylie Everett, Sharlet Ramsland

Oil and gas and alternative energy have been active areas of patenting. The resulting patents represent both an opportunity to their holders and a litigation risk to others. Selected patents from these core technologies have been asserted in litigation by operating companies and non-practicing entities (NPEs) alike. However, oil and gas companies also face litigation risk beyond core technologies, especially as their operations incorporate non-core technologies that have been the subject of serial litigation, such as wireless communications, remote monitoring and control and real-time asset tracking. These sectors have been active areas of litigation, especially from NPE plaintiffs.

This presentation explores patenting in oil and gas and patent litigation both in core and non-core technologies, with a focus on NPEs.

32a How to Protect Your Intellectual Property in the Academic World

Paul Townsend, Charles Collins-Chase, Lauren Dowty

Universities are full of academics working on cutting edge technology, but how is all that innovation protected? Universities often have their own organizations in place to help encourage, protect, and license innovations occurring at the school. This presentation will provide an overview of the basics of patent law, how tech transfer works at a university, and important considerations for students, professors, and the university itself. Speakers will also highlight several case studies addressing whether a university's intellectual property protection was sufficient.

32b Obviously New or Just a Copy – Engineers Play a Critical Role in Evaluating Patent or Trade Secret Infringement Claims

Achim Wechsung, David Hietala

Innovation is a key driver for the future of the chemical industry. Inventors face the choice to protect their novel technologies using patents or keeping them trade secrets. The intent of a patent is to foster innovation by providing inventors an assurance that they hold exclusive rights to the invention which is legally protected for a defined period of time. However, the issuance of a patent also includes technical details of the invention and its merits, which become a matter of public record. As a result, patented technologies may be susceptible to infringement (unauthorized use, manufacture, sale, or importation), which can result in litigation. Trade secrets, on the other hand, are not disclosed publicly and must be actively protected, for example, using non-disclosure agreements. Similarly, litigation may arise out of trade secret disputes related to breach of contract, unauthorized disclosure, and validity of a trade secret claim.

Engineers play a critical role in assessing claims in patent and trade secret litigation either as a technical consultant, or a testifying expert witness. Engineers can help legal professionals analyze and understand the technical claims in a patent or trade secret, which oftentimes, are related to very complex, novel, and complicated technologies. Engineers can evaluate patent validity through review of prior art, the claims in prior patents, and the state of scientific knowledge at the time a patent was issued. Engineers also provide insights into if trade secrets could have been obtained independently. Additionally, engineers can also help design experiments and/or analyses to test and validate a patent's technical claims. Lastly, engineers can provide supporting analysis that helps demonstrate the novelty, or lack of novelty, of a patent or trade secret.

In this presentation, we review the application of chemical engineering principles as a tool to help evaluate patent infringement or trade secret breach claims in intellectual property litigation. Three case studies will be used to illustrate key concepts. First, an example involving a company that raised questions about the technical basis for claims in a patent filed by an industry competitor will be presented. Second, a case study involving the misappropriation of a trade secret for removing fuel contaminants specifying reactors and their operating conditions will be discussed. And lastly, a case study surrounding a defendant's use of a patented catalyst used to manufacture an industrial chemical will be discussed.

AUTHOR INDEX

NOTES

NOTES

NOTES

NOTES

NOTES

NOTES

NOTES